Siomara Dias da Costa Lemos
Leandro V. Astarita

Melissa officinalis and the production of substances of medicinal interest

Siomara Dias da Costa Lemos
Leandro V. Astarita

Melissa officinalis and the production of substances of medicinal interest

A literature review on plant response to biotic and abiotic stresses

ScienciaScripts

Imprint

Any brand names and product names mentioned in this book are subject to trademark, brand or patent protection and are trademarks or registered trademarks of their respective holders. The use of brand names, product names, common names, trade names, product descriptions etc. even without a particular marking in this work is in no way to be construed to mean that such names may be regarded as unrestricted in respect of trademark and brand protection legislation and could thus be used by anyone.

Cover image: www.ingimage.com

This book is a translation from the original published under ISBN 978-613-9-65541-0.

Publisher:
Sciencia Scripts
is a trademark of
Dodo Books Indian Ocean Ltd. and OmniScriptum S.R.L publishing group

120 High Road, East Finchley, London, N2 9ED, United Kingdom
Str. Armeneasca 28/1, office 1, Chisinau MD-2012, Republic of Moldova, Europe
Printed at: see last page
ISBN: 978-620-7-78052-5

SUMMARY

CHAPTER 1

Herbal medicines

Known from primitive peoples to the present day, plants with medicinal activity have traditionally been used for food, disease prevention and treatment (Yunes and Calixto 2001). Medicinal plants, from trees to herbs, are capable of producing biologically active molecules through their secondary metabolism. These molecules, present in plant extracts, are considered herbal medicines (Yunes et al. 2001).

Natural products of plant origin have been used both in medical therapy and in the cosmetics and food industries, serving as flavorings and antioxidants. Most of the natural products used medicinally are terpenoids, quinones, lignans, flavonoids and alkaloids and, in general, have tranquilizing, analgesic, anti-inflammatory, cytotoxic, contraceptive, antimicrobial, antiviral and fungicidal actions, among others (Phillipson 2001).

The use of herbal medicines offers numerous advantages for human beings, especially the high degree of trust that many segments of the population have in them. Thus, studies suggest that herbal medicines may have fewer unpleasant side effects than synthetic drugs, since the active compounds are presented in reduced concentrations (Schulz 2002).

Several authors have pointed out that compounds with biological activity that are similar to drugs can act on different molecular targets (Yunes et al. 2001;

Darshan and Doreswamy 2004; Müller et al. 2004; Butterweck et al. 2004; Silano et al. 2004). In addition, the development of new herbal medicines and the costs of researching them can be lower than those used to manufacture pharmaceuticals (Yunes et al. 2001).

Melissa officinalis

Melissa officinalis L. (melissa) belongs to the Lamiaceae family, subfamily Nepetoidae (Janicsàk et al. 1999). Its main taxonomic feature is the presence of a phenolic acid called rosmaric acid, an ester of caffeic acid (Wang et al. 2004). Phenolic compounds and essential oils can also be found in its composition (Janicsàk et al. 1999). In popular culture, the dried and fresh leaves of this plant are recommended for anxiety, insomnia, colds, digestive problems and rheumatic pains (Cuppett 1998; Lorenzi and Matos 2002).

The essential oils of this plant are characterized by a mixture of volatile monoterpenes and sesquiterpenes (groups of terpenes insoluble in water and synthesized from acetyl CoA or glycoside intermediates biosynthesized from primary metabolism), with an oil-like consistency, insoluble in water and soluble in organic solvents, which give it a characteristic aroma or taste (Wachowicz and Carvalho 2002).

Carnat et al. (1998), working with *M. officinalis* essential oils, observed that among the phenylpropanoids, caffeic acid and rosmaric acid, as well as some flavonoids, were the most abundant compounds.

Essential oils can be found in specialized plant structures, such as glandular trichomes (Wachowicz and Carvalho 2002), and have insect and herbivore repellent properties, as well as being used in the cosmetics and food industries (Li et al. 2005). These oils can also have antimicrobial activities (Moreira et al. 2005) and act as modulators on certain enzymes, such as acetylcholinesterase (Salah and Jager 2005), presenting potential interest in the pharmaceutical field.

Mrlionovà et al. (2002) point out that there are variations in the constitutions of essential oils in the various species of Lamiaceae, indicating that there is an increase in the production of oils and phenylpropanoids, especially after the flowering period (Badi et al. 2004). In this sense, growing *M. officinalis* in saline soils results in a significant reduction in the growth and synthesis of essential oils (Ozturk et al. 2004), and the type of substrate and the time of harvesting the aerial parts can also result in alterations to the oil constituents (Manukyan et al. 2004).

In addition to oils, the presence of phenolic compounds in plants of the Lamiaceae family (Wang et al. 2004; Kovatcheva et al. 1996) and Boraginaceae (Yamamoto et al. 2000) provides an effective defense against herbivores and fungi, as well as pigmentation (Del Bano et al. 2004a).

However, like essential oils, the presence of phenylpropanoids and their activities depend on the quality of the plant, geographical origin, climatic conditions to which the plant is exposed, harvesting period and storage methods (Dixon and Paiva 1995; Areias et al. 2000; Santos-Gomes et al.

2002).

Botanical characteristics

M. officinalis is characterized by being a perennial herbaceous plant, branched from the base and erect, reaching a height of 30-60 cm. It is a plant native to Europe that is now cultivated in several countries. The leaves are membranous and wrinkled and approximately 4 cm long. The flowers are slightly whitish in color and are propagated by seeds or cuttings (Lorenzi and Matos 2002).

CHAPTER 2

Secondary Metabolites in Plants

Throughout the plant kingdom, the cell's essential metabolism is carried out by what is considered primary metabolism, i.e. the metabolism responsible for cell growth and functioning. This metabolism produces substances such as amino acids, proteins, carbohydrates, lipids and nucleic acids, with the function of maintaining cell structures, storing and using energy (Serafini et al. 2001; Maraschin and Verporte 1999; Taiz and Zeiger 2004).

Unlike primary metabolism, plants have a diversity of biologically active molecules, called secondary metabolites, which are not common to all plants and which characterize certain taxonomic groups (Chen and Chen 2000). These metabolites are not directly involved in the plant's growth and development, but are responsible for the plant's survival in its environment (Serafini and Barros 2001; Taiz and Zeiger 2004), promoting protection against herbivory and infection by pathogenic microorganisms, as well as attracting pollinating animals and acting as allelochemicals in plant-plant competition, ensuring the survival of the species in the ecosystem (Maraschin and Verporte 1999). These metabolites may also have biological activities at low concentrations and be related to aromas, flavors and natural dyes (Brown et al. 1989), varying according to the development of the plant.

During the flowering of *Rosmarinus officinalis* there is an increase in the levels of carnosic acid (a diterpenic phenolic compound) in the leaves and flowers (Del Bano et al. 2004a), while the concentration of flavonoids in the leaf stems increases during leaf development (Del Bano et al. 2004b). In this way, the concentration of certain secondary metabolites may be related to the stages of plant development, with the accumulation of different compounds due to processes of synthesis and transport in the different stages of plant development. Among the secondary metabolites produced by plants, the phenylpropanoids represent one of the most important groups that perform protective activity in plants, preventing the attack of pathogens, herbivory and parasitism   , mainly under conditions of conditions stress (Dixon and Paiva 1995).

The use of modern techniques such as genetic engineering and molecular biology represents an advance in the knowledge and manipulation of the compounds produced in secondary metabolism. In this way, it is possible to alter the quantity of certain groups of compounds (Verpoorte and Memelink 2002), such as flavonoids in glandular trichomes, which act to reduce herbivory in *Medicago sativa* (Aziz et al. 2005), clone important enzymes in the phenylpropanoid biosynthesis pathway (Yu et al. 2001) and modifying the metabolism of cysteine (Riemenschneider et al. 2005), as well as using molecular markers to select medicinal plants with lower levels of toxic components (Canter et al. 2005).

Phenylpropanoids

Phenylpropanoids, also known as phenolic compounds, are widely distributed in the plant kingdom (Kurkin 2003; Kosar et al. 2004) and have attracted the attention of researchers due to their potential use as natural antioxidants (Cuppett 1998). This group of molecules acts as free radical scavengers, such as superoxide O_2^- and reactive nitrogen species (Sokmen et al. 2005), preventing the peroxidation of lipids (Dixon and Paiva 1995), acting to prevent diseases that involve the action of free radicals and the oxidation of lipoproteins, such as cardiovascular problems, arteriosclerosis and thrombosis (Manach et al. 1998; Wang and Zhang 2005), as well as modulating physiological responses in animals, such as vasodilation and inflammation (Rice-Evans and Packer 2003).

Phenylpropanoids are characterized by having a phenol group (a functional hydroxyl group and an aromatic ring) and are produced by the secondary metabolism of plants where they perform a variety of functions (Simoes et al. 2000; Taiz and Zeiger 2004). This class of natural compounds is often conjugated to sugars and occasionally present in proteins, alkaloids and terpenes (Harborne 1988).

These compounds can be formed through two biogenetic routes (Figure 1): by shikimic acid, from carbohydrates, via the mevalonate pathway, which is initiated by acetyl-coenzyme A and malonyl-coenzyme A. This last route is the least significant because it is restricted to fungi and bacteria (Taiz and Zeiger 2004).

Figure 1: Main phenylpropanoid metabolism pathways and their interconnections (adapted from Taiz and Zeiger 2004)

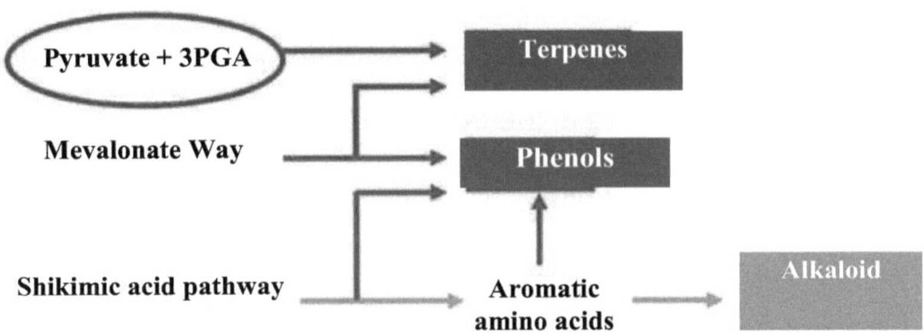

The phenylpropanoid pathway is considered to be one of the most important metabolic pathways responsible for the synthesis of many natural products in plants, such as flavonoids and anthocyanins, tannins, lignins and phenolic acids. The flavonoids synthesized are responsible for the color of flowers, fruits and some leaves, and also have the properties of inhibiting infections caused by pathogens, protecting the cell against UV radiation, inducing pollen germination and regulating auxin transport (Hao et al. 1996; Harborne and Williams 2000; Nugroho et al. 2002).

Phenolic acids have two different classes, hydroxybenzoic acids and hydroxycinnamic acids, derived from benzoic acid and cinnamic acid respectively (Fleuriet and Macheix 2003). Among the phenolic acids of greatest importance to humans, rosmaric acid and caffeic acid, present in

9

large quantities in individuals from the Lamiaceae family, have attracted the attention of researchers due to a series of biological activities such as antidepressant (Takeda et al. 2002), antioxidant (Akowuah et al. 2004) and antispasmodic (Darshan and Doreswamy 2004). These acids derived from the cinnamic acid pathway (Dixon and Paiva 1995), as well as having therapeutic potential, are also used in the treatment and prevention of various diseases, including bronchial asthma, ulcers, inflammation and hepatotoxicity (Al-Sereitia et al. 1999).

Caffeic acid (CA), which normally occurs in lower concentrations than rosmaric acid, can be found in various parts of the plant such as leaves, flowers and stems. However, it can also be present in the soil, inhibiting germination and plant growth (Banthorpe et al. 1989; Janicsàk et al. 1999), and together with lignin providing mechanical rigidity to plants, preventing their consumption and making it difficult for herbivores to digest them (Taiz and Zeiger 2004). One of its main properties is its antimicrobial effect against *Staphylococcus aureus* (Kurkin 2003).

Considered a neolignin, rosmaric acid (RA) can be used as a marker of authenticity and quality of *R. officinalis* and *M. officinalis* plants (Kurkin 2003). Structured as a disaccharide, this acid acts as a biosynthetic intermediate in the metabolism of phenylpropanoids in dicotyledons (Moolgaard and Ravn 1988).

Several studies indicate that RA has the property of reversing inflammatory processes (Molnar and Garai 2005), inhibiting the enzymatic activity of lipooxygenases, cyclooxygenases and xanthine oxidase (Chung et al. 2004), in addition to its antioxidant action (Al-Sereitia et al. 1999). Due to their antioxidant properties, phenolic compounds, including RA, can act as protectors against human cancer (Al-Sereitia et al. 1999; Kovatcheva et al. 1996).

RA synthesis can be promoted by manipulating biotic conditions, as in cultures of *Coleus blumei* (Petersen and Simmonds 2003; Banthorpe et al. 1989; Yamamoto et al. 2000). In this way, RA biosynthesis can be promoted by inducing plant-specific defense responses (Chen and Chen 2000; Petersen and Simmonds 2003).

Abiotic and biotic factors that interfere with secondary metabolism

The biosynthesis of secondary metabolites is closely related to environmental conditions. These conditions promote alterations both in the routes of synthesis and degradation of compounds and in genetic activity in response to some type of stress (Wilt and Miller 1992; Badi et al. 2004), promoting alterations in growth and in the quantity or quality of secondary compounds produced by plants (Badi et al. 2004).

Plant tolerance to stress involves adaptive responses related to molecular and cellular processes, resulting in damage limitation and disease prevention (Tester and Bacic 2005). Studies carried out on *Glycine max* indicate that abiotic conditions, such as low temperature, pH and high salinity, promote changes in the transcript levels of nodulation genes (*nod genes*), leading to root deformation (Duzan et al. 2004). In maize, both cold stress and phosphate deficiency in the soil promote the synthesis and accumulation of anthocyanins (Christie et al. 1994), just as the time of year and the level of fertilization affect the quality of the essential oil in *Salvia officinalis* (Areias et al. 2000).

Extreme environmental conditions, such as salinity, aridity and extreme temperatures, can act as limiting factors for growth and productivity (Atienza et al. 2004). Tester and Bacic (2005) observed the effect of environmental conditions such as salinity and soil acidity on cereals, while Georgiev et al. (2004) described that exposing *in vitro* cultures of *Lavandula vera* to different temperatures promotes changes in the synthesis and accumulation of bioactive molecules.

Plants respond to abiotic and biotic stress conditions by altering the profile of secondary metabolites (Dixon and Paiva 1995), with phenylpropanoids being the group that shows the greatest changes (Randhir et al. 2002). Thus, phenolic acids are directly involved in responses to mechanical, chemical and biological stresses, and certain molecules from this group can be synthesized *de novo* in response to attack by pathogens (Fleuriet and Macheix 2003).

In this way, the phenylpropanoid pathway can be promoted using compounds (elicitors) that simulate adverse environmental conditions.

Elicitors

Many techniques have been developed to improve the productivity of compounds of interest in plants, including the selection of strains, optimization of growing conditions, the use of elicitors and metabolic engineering (Chen and Chen 2000). The use of elicitor molecules represents an efficient approach to increasing secondary metabolism. These molecules can be endogenous or exogenous, depending on how they act on the plant, and biotic or abiotic, depending on their origin (Dornenburg and Knorr 1995).

According to Pitta-Alvarez et al. (2000), extracts from various microorganisms (such as pathogenic bacteria and fungi) and molecules that signal plant defense mechanisms (such as salicylic acid, jasmonic acid, etc.), as well as enzymes (cellulase, pectinase, etc.), are considered biotic elicitors. According to the same authors, abiotic elicitors include ultraviolet radiation, heavy metals and chemical compounds.

The use of biotechnological techniques, such as *in vitro* tissue and cell culture, makes it possible to identify biotic and abiotic elicitors capable of stimulating the production of metabolites of agricultural, pharmaceutical and food interest (Altman 1999; Qin and Lan 2004). In addition to determining elicitor molecules, these techniques also make it possible to study the mechanisms of pathogen infection in plants (Puchooa 2004) and synthesis pathways, such as rosmaric acid (Petersen 1997).

Supplementation with elicitors such as methyl jasmonate and chitosan induces an increase in paclitaxel synthesis in *Taxus chinensis* cell cultures (Luo and He 2004). Pot plants of *Hypericum perforatum* inoculated with fungal spores double the production of hypericin, an anthraquinone with pharmacological action (Sirvent and Gibson 2002), and there are several studies reporting an increase in the synthesis and accumulation of compounds of interest in response to the action of elicitors, such as the production of berberine (Brodelius et al. 1989), taxol (Ciddi et al. 1995; Dornenburg and Knorr 1995) and new umbelliferones in *Ammi majus* L. (Staniszewska et al. 2003).

Pathogens as elicitors

The invasion of an intact plant by pathogens activates numerous plant defense mechanisms, including the synthesis of antimicrobial metabolites (Dornenburg and Knorr 1995).

In general, plants naturally defend themselves against pathogens using a combination of two strategies:

(i)   structural, developing physical barriers that prevent both the entry of the pathogen and its spread through the plant's tissues;

(ii)   biochemical reactions of cells and tissues, producing toxic compounds that prevent the pathogen from developing in the plant (Medeiros et al. 2003). Local biochemical reactions, which lead to localized cell death and the formation of necrotic lesions at the site of pathogen penetration, are considered hypersensitivity responses (HR) (Sticher et al. 1997).

However, either HR or another response mechanism to pathogen attack can induce long-lasting, broad-spectrum resistance in the plant to subsequent infections. This induced disease resistance response has been known for many years as Systemic Acquired Resistance (SAR) (Ryals et al. 1994). The induction of SAR can occur after the treatment of plants with substances of both biotic and abiotic origin, and may or may not be accompanied by the accumulation of molecular markers, such as salicylic acid (Curtis et al. 2004).

Flavonoids, coumarins and phenolic acids such as caffeic, chlorogenic and ferulic acids are chemical defenses of plants. These compounds can be present constitutively or be synthesized de *novo* due to the presence of pathogens (Friedmam 1997). These molecules, which were not present before the infection, are called phytoalexins and have antimicrobial properties.

According to Staniszewska et al. (2003), it is possible to promote the biosynthesis of secondary compounds, such as umbelliferone, *in Ammi majus* roots grown *in vitro*, elicited with dead *Enterobacter sakazaki* cells. Dietrich et al. (2005) describe that exogenous application of the bacterium *Pseudomonas syringae* to *Arabidopsis sp.* promotes systemic responses resulting in increased peroxidase and chitinase activity in infected leaves.

Likewise, the exogenous application of phytopathogenic organisms to plants can trigger an increase in the synthesis of molecules of interest, such as rosmaric acid in *Colleus* sp. (Petersen and Simmonds 2003), production of dopamine precursors in *Vicia faba* (Randhir et al. 2002), production of phenylpropanoids in *Brassica napus L. var. oleifera* (Plazek et al. 2005), oxidizing compounds in *Triticum aestivum* (Ortmann et al. 2004) and scopolamine and hyoscyamine in *Brugmansia candida* (Pitta-Alvarez et al. 2000).

Among the phytobacteria of interest for eliciting plants in order to induce SAR, *Erwinia carotovora* stands out. As a gram-negative bacterium, it is the most severe bacterium to attack *Solanum tuberosum* and can cause symptoms in various parts of the plant, such as leaves, roots, stems, tissues and tubers. When attacked by this bacterium, the accumulation of phytoalexins is considered to be one of the most important defense mechanisms and is related to reducing the susceptibility of plants to this bacterium (Abenthum et al. 1995).

Action of Salicylic Acid (AS)

The exogenous application of salicylic acid (AS) to plants triggers biosynthetic pathways for growth and development, as well as the expression of defense genes that are activated when pathogen infection occurs. Both AS and methyl jasmonate are involved in signal transduction systems that induce the production of defense compounds such as alkaloids, polyphenols and pathogenesis-related proteins. This response results in the plant's defense, protection and resistance against pathogens (Yao and Tian 2005).

Salicylic acid (AS) can induce flowering in Lamiaceae (Hatayama and Takeno

2003), protect *Triticum aestivum* against water stress (Singh and Usha 2003) and promote HR, which is essential for local and systemic resistance against pathogens (Pastirovà et al. 2004). AS and its chemical derivatives, such as acetylsalicylic acid (ASA), can also act as tropane metabolism inducing agents in *Atropa belladonna* cells and tissues grown *in vitro* (Lee et al. 2001), anthraquinones in *Morinda citrifolia* (Komaraiah et al. 2005), modulate the activity of glutathione-S-transferase (Gong et al. 2005), and promote the activity of phenylalanine ammonia, peroxidases and lignification in potted plants of *Asparagus officinalis* (He and Wolyn 2005).

AS can regulate the flavonoid formation pathway and is considered by some authors to be a phytohormone involved in plant defense reactions (Gutiérrez-Conrado et al. 1998; Nugroho et al. 2002), inducing the systemic acquired response (SAR) (Gutiérrez-Conrado et al. 1998; Nugroho et al. 2002; Verpoorte and Memelink 2002; Curtis et al. 2004). In addition to triggering these responses, AS is also involved in the activation of genes related to stress responses to drought, cold, heat, salinity and UV radiation (Peng and Jiang 2006), acting at the level of ethylene biosynthesis, membrane depolarization, an increase in the photosynthetic pathway and the amount of chlorophyll in some plants, such as soybeans (Gutiérrez-Conrado et al. 1998).

The exogenous application of AS can reduce plant growth. Gutiérrez-Conrado et al. (1998) describe that different concentrations of AS can result in a 20% to 23% reduction in the growth of *Glycine max,* with the most dramatic effect being observed in the roots, which were reduced by approximately 45%.

There is a close relationship between the activity of phenylalanine ammonia lyase (PAL) and AS biosynthesis (Figure 2), and an increase in the activity of this enzyme can promote the inhibition of AS signaling (Pastirovà et al. 2004; Peng and Jiang 2006; Yao and Tian 2005).

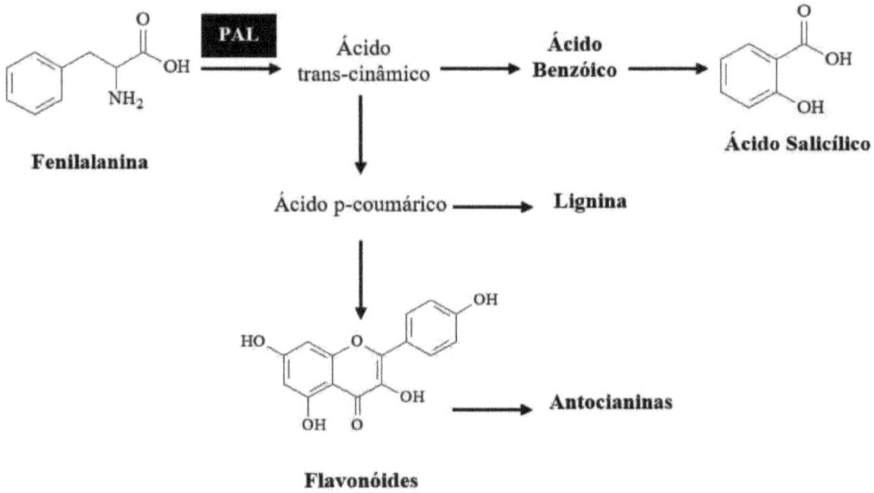

Figura 2: Synthesis route for phenolic compounds derived from the phenylalanine ammonia lyase (PAL) pathway (adapted from Taiz and Zeiger 2004).

Phenylalanine ammonia lyase (PAL) activity

The induction of stress in plant organisms by biotic and abiotic factors results in defense responses. The production of defense molecules, such as anthocyanins, flavonoids, condensed tannins and isoflavonoids, represents the so-called HR reaction, which can lead to the localized death of plant cells located where the aggressor penetrates (Pastirovà et al. 2004).

HR is triggered by enzymes of the phenylpropanoid pathway, activated mainly by phenylalanine ammonia lyase (PAL), the first and most important enzyme involved in this pathway (Figure 2) (Pastirovà et al. 2004; Yao and Tian 2005). PAL is an enzyme that uses L-phenylalanine as a substrate to produce trans-cinnamic acid (Hao et al. 1996).

Being related to the plant's defense system, PAL activity can be induced by the exogenous supplementation of elicitors such as salicylic acid and methyl jasmonate (Yao and Tian 2005), resulting in an increase in the production of phenolic compounds (Taiz and Zeiger 2004). In this sense, virus attack on legumes promotes a rapid increase in PAL activity, which is closely related to the physiological state of the plant (Nugroho et al. 2002).

In addition to PAL, the activity of the enzyme chalcone synthase (CHS) can also be promoted by stress conditions. CHS is a key enzyme in flavonoid synthesis (Figure 3), forming anthocyanins, flavonoids, condensed tannins and isoflavonoids. Flavonoids, such as quercetin, and flavones are important not only as anti-pathogenic agents, but also for protecting plants against UVB radiation (280-320 nm).

Arabidopsis thaliana mutants that do not express this enzyme are more sensitive to UVB radiation and grow poorly under normal conditions due to the lack of flavonoid production (Dixon and Paiva 1995).

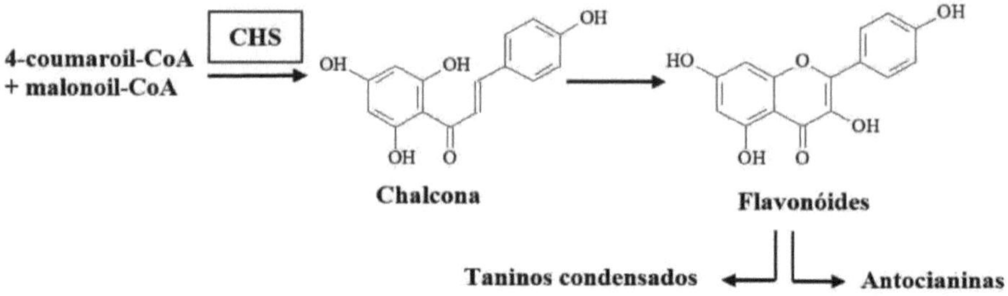

Figura 3: Flavonoid synthesis pathway, in which chalcone synthase (CHS) is the main enzyme involved in the biosynthesis of anthocyanins, isoflavonoids and tannins (adapted from Taiz and Zeiger 2004).

Polyphenol oxidase (PPO) activity

Polyphenol oxidase (PPO) is an enzyme widely distributed in nature, known to be part of a group of enzymes that are related to the browning of damaged fruits and vegetables (Gauillard et al. 1993), as well as melanization in animals (Shi et al. 2001). This group of enzymes catalyzes two distinct events related to molecular oxygen, called (a) the hydroxylation of monophenols into diphenols, and (b) the subsequent oxidation of diphenols into quinones (Valero and Garcia-Carmona 1992; Gauillard et al. 1993; Shi et al. 2001).

Phenolic compounds, located in vacuoles, are intermediates in the metabolism of phenylpropanoids and are involved in the formation of cell walls, tissues and plant organs. Precursors for lignin synthesis, their deposition in the cell wall after pathogen infection is an important plant defense mechanism. On the other hand, PPO is localized in the chloroplasts, in the membranes of the thylakoids, and in some species, it can present a latent form (Valero and Garcia-Carmona 1992). When the plant tissue is damaged, the oxidation of these phenolic compounds by PPO begins (Felton et al. 1992). Cvikrova et al. (1996) observed in alfalfa that the maximum activity of PPO was related to the presence of free phenolic compounds.

In order to clarify its various biological functions, many routes have been proposed. However, authors have suggested several functions of this enzyme related to plant resistance against pathologies (Constabel et al. 1995; Ray and Hammerschmidt 1998), as well as against herbivory and insects (Felton et al. 1992). In some plants, the increase in PPO activity is related to the production of secondary compounds, such as jasmonic acid and methyl jasmonate, involved in the transduction of defense signals.

However, this fact is not generalized in plants, as some plants show little or no induction of this enzyme when injury or elicitation occurs (Constabel et al. 1995). The relationship between PPO and phenolics is made clearer by the work of Zheng and Shetty (2000), who describe that there is a close relationship between PPO and phenolic compounds, since this oxidation is necessary to convert these free compounds into the form of lignin in situations of pathogen attack.

Mazzafera and Robinson (2000), working with *Coffea arabica,* described an increase in PPO activity in response to pathogen attack, inducing anti-nutritive defense. In this sense, during wounding of the plant, quinones are formed by the oxidation of phenols induced by PPO, which can modify plant proteins, reducing their nutritional value for herbivores. These same authors report that young leaves of this plant show high activity of this enzyme, and that this activity declines concomitantly with leaf development.

CHAPTER 3

Micropropagation

Among the various biotechnological techniques used on plants, micropropagation stands out due to its potential use in controlling plant growth and development, enabling in-depth studies in the biochemical, pharmaceutical and food areas, by controlling the synthesis of natural products.

Micropropagation is a technique that does not require much technology and belongs to the category of low-cost techniques (Puchooa 2004). Using this technique, a large number of high-quality plant clones can be obtained and new genotypes can be propagated quickly and on a large scale from a small sample of germplasm (Altman 1999).

It is extremely difficult to compare the amount of a substance produced *in* cultured cells and tissues with those produced by the plant *in vivo,* as some metabolites are often synthesized in one tissue and transported to others, where they are accumulated. Despite this limitation, there are advantages to using tissue culture as a source of biologically important compounds, such as independence from environmental factors, including climate, pests, diseases and geographical limitations (Mantell et al. 1994).

There are several protocols for the regeneration of various plant species of interest, based on variations and combinations of growth regulators supplemented to the culture medium (Djilianov et al. 2005).

The techniques used in tissue culture allow for the production of various molecules of interest, such as flavorings and dyes used in industry (Millam et al. 2005). Because of this capacity, plants have been genetically modified to produce more complex molecules, such as antibodies or vaccines, opening up a new era for molecular agriculture (Puchooa 2004).

With the aim of producing uniform seedlings with a known genetic identity, the culture of nodes and stem apices (Figure 4) makes it possible to obtain a large number of identical individuals from selected plants (homogeneity), produce seedlings all year round and develop techniques for germplasm conservation and genetic improvement (Serafini et al. 2001).

Figure 4: Stages of micropropagation (adapted from George 1993).

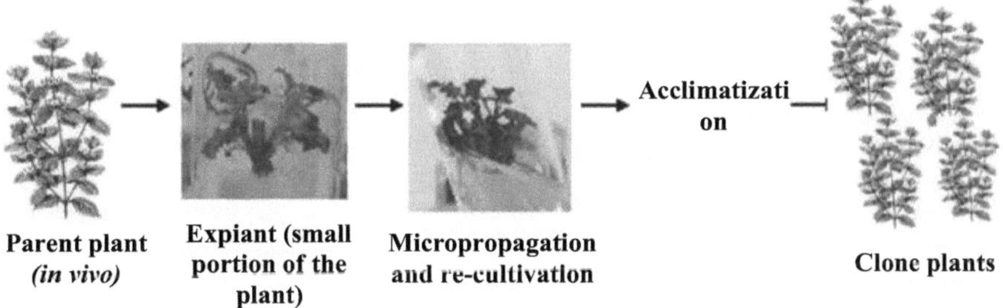

Induction of Secondary Metabolites

The production of secondary metabolites *in vitro* has great potential for understanding biosynthesis processes. This is because cell cultures can be subjected to different physical and chemical conditions, such as the use of precursors, elicitors, stresses and blockers of synthesis pathways, leading to changes in the production of a given metabolite (Djilianov et al. 2005).

Many studies are being carried out on the tissue culture of various plant species of commercial interest. Some of these, carried out with *Salvia officinalis, have made* it possible to study the effect of the concentration of kinetin on the production of phenolic diterpenes, mainly carnosol (Santos-Gomes et al. 2002). According to Staniszewska et al. (2003), the use of elicitors such as fungi and silica promotes the synthesis of secondary metabolites *in Ammi majus* cells and roots grown *in vitro*. In European countries, this alternative to manipulation and cultivation stands out, since the acclimatization of certain plant species in these regions is difficult (Staniszewska et al. 2003).

In addition to these studies, Amaral and Silva (2003) highlight work on the production of *Aloe vera* clones, the production of hybrids by somatic hybridization in *Nicotiana tabacum*, haploids in *Hyosciamus niger* and vaccine-generating plants in *Solanum tuberosum*.

The tissue culture technique also makes it possible to use elicitors (chemical agents and stressors) to alter metabolic pathways, affecting the quality and quantity of the bioactive molecules produced. In this sense, work with *Atropa belladona* and *Lithospermum erythrorhizon* has shown good results in terms of the production of metabolites with therapeutic properties (Lee et al. 2001, Yamamoto et al. 2000).

BIBLIOGRAPHICAL REFERENCES

Abenthum, K.; Hildenbrand, S.; Ninnemann, H. 1995. Elicitation and accumulation of phytoalexin in stems, stolons and roots of *Erwinia-infected* potato plants. **Physiological and Molecular Plant Pathology 46**: 349-359.

Amaral, C.L.F. and Silva, A.B. 2003. Biotechnological improvement in medicinal plants. **Biotecnologia Ciência e Desenvolvimento 30**: 55-59.

Akowuah, G.A.; Zhari, I.; Norhayati, I.; Sadikun, A.; Khamsah, S.M. 2004. Sinensetin, eupatorin, 30-hydroxy-5, 6,   7, 40-tetramethoxyflavone and rosmarinic acid contents and antioxidative effect of Orthosiphon stamineus from Malaysia. **Food Chemistry 87: 559-566.**

Al-Sereitia, M.R.; Abu-Amerb, K.M.; Sena, P. 1999. Pharmacology of rosemary (*Rosmarinus officinalis* Linn.) and its therapeutic potentials. **Indian Journal of Experimental Biology 37:** 124-131.

Altman, A. 1999. [st]Plant biotechnology in the 21st century: the challenges ahead. **Electronic Journal of Biotechnology 2** (2): 51-55.

Areias, F.; Valentâo, P.; Andrade, P.B.; Ferreres, F.; Seabra, R.M. 2000. Flavonoids and phenolic acids of sage: Influence of some agricultural factors. **Journal Agriculture Food Chemistry 48:** 6081-6084.

Atienza, S.G.; Faccioli, P.; Perrotta, G.; Dalfino, G.; Zschiesche, W.; Humbeck, K.; Stanca, M. A.; Cattivelli, L. 2004. Large scale analysis of transcripts abundance in barley subjected to several single and combined abiotic stress conditions. **Plant Science 167:** 1359-1365.

Aziz, N.; Paiva, N.L.; May, G.D.; Dixon, R.A. 2005. Transcriptome analysis of alfalfa glandular trichomes. **Planta 221:** 28-38.

Badi, H.N.; Yazdani, D.; Ali, S.M.; Nazari, F. 2004. Effects of spacing and harvesting time on herbage yield and quality/quantity of oil in thyme, *Thymus vulgaris* L. **Industrial Crops and Products 19:** 231-236.

Banthorpe, D.V.; Bilyard, H.J.; Brown, G.D. 1989. Enol Esters of caffeic acid in several genera of the labitae. **Phytochemistry 28** (88): 2109-2113.

Brodelius, P.; Funk, C.; Haner, A.; Villegas, M. 1989. A procedure for the determination of optimas chitosan concentrations for elicitation of cultured plant-cells. **Phytochemistry 28** (10): 2651-2654.

Brown, C.M.; Campbell, I.; Priest, F.G. 1989. **Introduction to biotechnology**. Zaragoza, Spain. Editora Acribia S.A. 167p.

Butterweck, V.; Derendorf, H.; Gaus, W.; Nahrstedt, A.; Schulz, V.; Unger, M. 2004. Pharmacokinetic herb-drug interactions - are preventive screenings necessary and appropriate? **Planta Medica 70:** 784-791.

Canter, P.H.; Thomas, H.; Ernst, E. 2005. Bringing medicinal plants into cultivation: opportunities and challenges for biotechnology. **Trends in Biotechnology 23** (4): 180-185.

Carnat, A.P.; Carnat, A.; Fraisse, D.; Lamaison, J.L. 1998. The aromatic and polyphenolic composition of lemon balm (*Melissa officinalis* L. subsp. *officinalis*) tea. **Pharmaceutica Acta Helveatiae 72:** 301-305.

Chen, H. and Chen, F. 2000. Effects of yeast elicitor on the grown and secondary metabolism of a high-tanshinone-producing line of the Ti transformed *Salvia miltiorrhiza* cells in suspension culture. **Process Biochemistry 35:** 837-840.

Christie, P. J.; Alfenito, M. R.; Walbot, V. 1994. Impact of low-temperature stress on general phenylpropanoid and anthocyanin pathways: Enhancement of transcript abundance and anthocyanin pigmentation in maize seedlings. **Planta 194:** 541-549.

Chung, T-W.; Moon, S-K.; Chang, Y-C.; Ko, J-H.; Lee,Y-C.; Cho, G.; Kim, SH.; Kim, J-G.; Kim, C-H. 2004. Novel and therapeutic effect of caffeic acid and caffeic acid phenyl ester on hepatocarcinoma cells: complete regression of hepatoma growth and metastasis by dual mechanism. **The FASEB Journal 18:** 1670-1681.

Ciddi, V.; Srinivasan, V.; Shuler, M.L. 1995. Elicitation of Taxus sp cell cultures for production of taxol. **Biotechnology Letters 17** (12): 1343-1346.

Constabel, C.P.; Bergey, D.R.; Rian, C.A. 1995. Systemin activates synthesis of wound-inducible tomato leaf polyphenol oxidase via the octadecanoid defense signaling pathway. **Plant Biology 92:** 407-411.

Cuppett, S. L. 1998. Antioxidant activity of Lamiaceae. **Advances in Food and Nutrition Research 42:** 245-271.

Curtis, H.; Noll, U.; Stormann, J.; Slusarenko, A.J. 2004. Broad-spectrum activity of the volatile phytoanticipin allicin in extracts of garlic (*Allium sativum* L.) against plant pathogenic bacteria, fungi and Oomycetes. **Physiological and Molecular Plant Pathology 65** (2): 79-89.

Cvikrova, M.; Hrubcova, M.; Eder, J.; Binarova, P. 1996. Changes in the levels of endogenous phenolics, aromatic monoamines, phenylalanine ammonia-lyase, peroxidase and auxin oxidase activities during initiation of alfalfa embryogenic and non-embryogenic calli. **Plant Physiology Biochemistry 34:** 853-861.

Darshan, S. and Doreswamy, R. 2004. Patented anti-inflammatory plant drug development from traditional medicine. **Phytotherapy Research 18:** 343357.

Del Bano, M.J.; Lorente, J.; Castillo, J.; Benavente-Garcia, O.; Del Rio, J.A.; Ortuno, A.; Quirin, K-W.; Gerard, D. 2004a. Phenolic diterpenes, flovones, and rosmarinic acid distribution during the development of leaves, flower, stems, and roots of *Rosmarinus officinalis*. Antioxidant activity. **Journal of Agricultural and Food Chemistry 51:** 4247-4253.

Del Bano, M.J.; Lorente, J.; Castillo, J.; Benavente-Garcia, O.; Marin, M.P.; Del Rio, J.A.; Ortuno, A.; Ibarra, I. 2004b. Flavonoid distribution during the development of leaves, flower, stems, and roots of *Rosmarinus officinalis*.

Postulation of a biosynthetic pathway. **Journal of Agricultural and Food Chemistry 52:** 4987-4992.

Dietrich, R.; Ploss, K.; Heil, M. 2005. Growth responses and fitness costs after induction of pathogen resistance depend on environmental conditions. **Plant, Cell and Environment 28:** 211-222.

Dixon, R. A. and Paiva, N. L. 1995. Stress-indeuced phenylpropanoid metabolism. **The Plant Cell 7:** 1085-1097.

Djilianov, D.; Genova, G.; Parvanova, D.; Zapryanova, N.; Konstantinova, T.; Atanassov, A. 2005. *In vitro* culture of the resurrection plant *Haberlea rhodopensis*. **Plant Cell, Tissue and Organ Culture 80:** 115-118.

Dornenburg, H. and Knorr, D. 1995. Strategies for the improvement of secondary metabolite production in plant-cell cultures. **Enzyme and Microbial Technology 17** (8): 674-684.

Duzan, H.M.; Zhou, X.; Souleimanov, A.; Smith, D.L. 2004. Perception of *Bradyrhizobium japonicum* Nod factor by soybean [*Glycine max* (L.) Merr.] root hairs under abiotic stress conditions. **Journal of Experimental Botany 55** (408): 2641-2646.

Felton, G.W.; Donato, K.K.; Broadway, R.M.; Duffey, S.S. 1992. Impact of oxidized plant phenolics on the nutritional quality of dietary protein to a noctuid herbivore, *Spodoptera exigua*. **Journal of Insect Physiology 38:**

277-285.

Fleuriet, A. and Macheix, J-J. 2003. Phenolic acids in fruits and vegetables. 2003. *In:* Rice-Evans, C. A.; Parcker, L. **Flavonoids in health and disease**. Marcel Dekker. Inc. New York. pp.1-41.

Friedmam, M. 1997. Chemistry, biochemistry, and dietary role of potato polyphenols: a review. **Journal Agriculture and Food Chemistry 45:** 15231540.

Gauillard, F.; Richard-Forget, F.; Nicolas, J. 1993. New spectrophotometric assay for polyphenol oxidase activity. **Analytical Biochemistry 215:** 59-65.

George, E.F. 1993. **Plant Propagation by Tissue culture: part 1 - the technology**. England. Ed. Exegetics Ltd. 574p.

Georgiev, M.; Pavlov, A.; Ilieva, M. 2004. Rosmarinic acid production by *Lavandula vera* MM cell suspension - the effect of temperature. **Biotechnology Letters 26:** 855-856

Gong, H.; Jiao, Y.; Hu, W-w; Pua, E-C. 2005. Expression of glutathione-S-transferase and its role in plant growth and development in vivo and shoot morphogenesis in vitro. **Plant Molecular Biology 57:** 53-66.

Gutiérrez-Conrado, M.A.; Trejo-Lopes, C.; Larque-Saavedra, A. 1998. Effects of salicylic acid on the growth of roots and shoots in soybean. **Plant Physiology Biochemistry 36** (8): 563-565.

Hao, Z.; Charles, D.J.; Yu, L.; Somin, J.E. 1996. Purification and characterization of phenylalanine ammonia-lyase from *Ocium basilicum*. **Phytochemistry 43** (4): 735-739.

Harborne, J.B. 1988. **Phytochemical methods. A guide to modern techniques of plant analysis**. Chapman and Hall. London. 302p.

Harborne, J.B. and Williams, C.A. 2000. Advances in flavonoid research since 1992. **Phytochemistry 55:** 481-504.

Hatayama, T. and Takeno, K. 2003. The metabolic pathway of salicylic acid rather than of chlorogenic acid is involved in the stress-induced flowering of *Pharbitis nil*. **Journal of Plant Physiology 160:** 461-467.

He, C.Y. and Wolyn, D. J. 2005. Potential role for salicylic acid in induced resistance of asparagus roots to *Fusarium oxysporum* f.sp. *asparagi*. **Plant Pathology 54:** 227-232.

Janicsàk, G.; Mathé, I.; Miklóssy-Vàri, V.; Blunden, G. 1999. Comparative studies of the rosmarinic and caffeic acid contents of Lamiaceae species. **Biochemical Systematics and Ecology 27:** 733-738.

Komaraiah, P.; Kavi Kishor, P.B.; Carlsson, M.; Magnusson, K-E.; Mandenius, C-F. 2005. Enhancement of anthraquinone accumulation in *Morinda citrifolia* suspension cultures. **Plant Science 168:** 1337-1344.

Kosar, M.; Dorman, D.; Baser, K.; Hiltunen, R. 2004. An improved HPLC post-column methodology for the identification of free radical scavenging phytochemicals in complex mixtures. **Chromatographia 60:** 635-638.

Kovatcheva, E.; Pavlov, A.; Koleva, I.; Ilieva, M.; Mihneva, M. 1996. Rosmarinic acid from *Lavandula vera* MM cell culture. **Phytochemistry 43** (6): 1243-1244.

Kurkin, V.A. 2003. Phenylpropanoids from medicinal plants: distribution, classification, structural analysis, and biological activity. **Chemistry of Natural Compounds 39** (2): 123-153.

Lee, K.; Hirano, H.; Yamakawa, T.; Kodama, T.; Igarashi, Y.; Shimomura, K. 2001. Responses of transformed root culture of *Atropa belladona* to salicylic acid stress. **Journal of Bioscience and Bioengineering 91** (6): 586-589.

Li, W.; Koike, K.; Asada, Y.; Yoshikawa, T.; Nikaido, T. 2005. Rosmarinic acid production by *Coleus forskohlii* hairly root cultures. **Plant Cell, Tissue and Organ Culture 80:** 151-155.

Lorenzi, H. and Matos, F. J. A. 2002. Medicinal plants of Brazil: native and exotic cultivated. **Nova Odessa**, SP. Instituto Plantarum, 511p.

Luo, J. and He, G. 2004. Optimization of elicitors and precursors for paclitaxel production in cell suspension culture of *Taxus chinensis* in the presence of nutrient feeding. **Process Biochemistry 39:** 1073-1079.

Manach, C.; Morand, C.; Crespy, V.; Demigné, C.; Texier, O.; Régérat, F.;

Rémésy, C. 1998. Quercetin is recovered in human plasma as conjugated derivatives which retain antioxidant properties. **FEBS Letters 426:** 331-336.

Mantell, S.H.; Matthaews, J.A.; McKee, R.A. 1994. **Principles of plant biotechnology - an introduction to genetic engineering in plants.** Ribeirao Preto. Brazilian Society of Genetics. 333p.

Manukyan, A.E.; Heuberger, H.T.; Schnitzler, A.H. 2004. Yield and quality of some herbs of the Lamiaceae family under soilless greenhouse production. **Journal of Applied Botany and Food Quality-Angewandte Botanik 78** (3): 193-199.

Maraschin, M. and Verporte, R. 1999. Secondary Metabolism Engineering. **Biotechnology Science and Development 10:** 24-28.

Mazzafera, P. and Robinson, S.P. 2000. Characterization of polyphenol oxidase in coffee. **Phytochemistry 55:** 285-296.

Medeiros, R. B.; Ferreira, M. A. S.V.; Dianese, J.C. 2003. **Mechanisms of aggression and defense in plant-pathogen interactions**. Ed. UnB, Brasilia.

Millam, S.; Obert, B.; Pret'ovà, A. 2005. Plant cell and biotechnology studies in *Linum usitatissimum* - a review. **Plant Cell, Tissue and Organ Culture 82:** 93-103.

Moolgaard, P. and Ravn, H. 1988. Evolutionary aspects of caffeoyl ester distribution in dicotyledons. **Phytochemistry 27** (8): 2411-2421.

Molnar, V. and Garai, J. 2005. Plant-derived anti-inflammatory compounds affect MIF tautomerase activity. **International Immunopharmacology 5:** 849-856.

Moreira, M.R.; Ponce, A.G.; del Valle, C.E.; Roura, S.I. 2005. Inhibitory parameters of essential oils to reduce a foodborne pathogen. **LWT 38:** 565570.

Mrlianovà, M.; Tekel'ova, D.; Felklova, M.; Reinohl, V.; Toth, J. 2002. The influence of the harvest cut height on the quality of the herbal drugs melissae folium and melissae herba. **Planta Medica 68:** 178-180.

Müller, R.S.; Breitkreutz, J.; Groning, R. 2004. Interactions between aqueous *Hypericum perforatum* extracts and drugs: *in vitro* studies. **Phytotherapy Research 18:** 1019-1023.

Nugroho, L.H.; Verberne, M.C.; Verpoorte, R. 2002. Activities of enzymes involved in the phenylpropanoid pathway in constitutively salicylic acid-producing tobacco plants. **Plant Physiology Biochemistry 40:** 755-760.

Ortmann, I.; Sumowski, G.; Bauknecht, H.; Moerschbacher, B. M. 2004. Establishment of a reliable protocol for the quantification of an oxidative burst in suspension-cultured wheat cells upon elicitation. **Physiological and Molecular Plant Pathology 64:** 227-232.

Ozturk, A.; Unlukara, A.; Ipek, A.; Gurbuz, B. 2004. Effects of salt stress and water deficit on plant growth and essential oil content of lemon balm. **Pakistan Jounal of Botany 36** (4): 787-792.

Pastirovà, A.; Repcàk, M.; Eliasovà, A. 2004. Salicylic acid induces changes of coumarin metabolites in *Matricaria chamomilla* L. **Plant Science 167:** 819824.

Peng, L. and Jiang, Y. 2006. Exogenous salicylic acid inhibits browning of freshcut Chinese water chestnut. **Food Chemistry** 94 (4): 535-540.

Petersen, M. 1997. Cytochrome P450: dependent hydroxylation in the biosynthesis of rosmarinic acid in *Coleus.* **Phytochemistry 45:** 1165-1172.

Petersen, M. and Simmonds, M.S.J. 2003. Rosmarinic acid. **Phytochemistry 62:** 121-125.

Phillipson, J.D. 2001. Phytochemistry and medicinal plants. **Phytochemistry 56** (3): 237-243.

Pitta-Alvarez, S.I.; Spollansky, T.C.; Giulietti, A.M. 2000. The influence of different biotic and abiotic alicitors on the production and profile of tropane alkaloids in hairly root cultures of *Brugmansia candida.* **Enzyme and Microbial Technology 26:** 252-258.

Plazek, A.; Hura, K.; Zur, I. 2005. Influence of chitosan, pectinase and fungal metabolites on activation of phenylpropanoid pathway and antioxidant activity in oilseed rape callus. **Acta Physiologia Plantarum 27** (1): 95-102 .

Puchooa, D. 2004. Biotechnology in Mauritius: current status and constraints. **Electronic Journal of Biotechnology 7** (2): 104-114.

Qin, W.M. and Lan, W.Z. 2004. Fungal elicitor-induced cell death in *Taxus chinensis* suspension cells is mediated by ethylene and polyamines. **Plant Science 166:** 989-995.

Randhir, R.; Shetty, P.; Shetty, K. 2002. L-DOPA and total phenolic stimulation in dark germinated fava bean in response to peptide and phytochemical elicitors. **Process Biochemistry 37:** 1247-1256.

Ray, H. and Hammerschmidt, R. 1998. Responses of potato tuber to infection by *Fusarium sambucinum*. **Physiology and Molecular Plant Pathology 53:** 81-92.

Rice-Evans, C.A. and Packer, L. 2003. **Flavonoids in health and disease**. Marcel Dekker, Inc. New York. 468p.

Riemenschneider, A.; Riedel, K.; Hoefgen, R.; Papenbrock, J.; Hesse, H. 2005. Impact of reduced *O-acetylserine*(thiol)lyase isoform contents on potato plant metabolism. **Plant Physiology 137:** 892-900.

Ryals, J.; Uknes, S.; Ward, E. 1994. Systemic Acquired Resistance. **Plant Physiology 104:** 1109-1112.

Salah, S.M. and Jager, A.K. 2005. Screening of traditionally used Lebanese herbs for neurological activities. **Journal of Ethnopharmacology 97:** 145-149.

Santos-Gomes, P.C.; Seabra, R.M.; Andrade, P.B.; Fernandes-Ferreira, M. 2002. Phenolic antioxidant compounds produced by *in vitro* shoots of sage (*Salvia officinalis* L.). **Plant Science 162:** 981-987.

Schulz, V. 2002. **Rational phytotherapy: a guide to phytotherapy for the health sciences**. Barueri. Ed. Manole. 386p.

Serafini, L.A.; Barros, N.M.; Azevedo, J.L. 2001. **Biotechnology in agriculture and agro-industry**. Guaiba. Ed. Agropecuària. 463 p.

Silano, M.; De Vincenzi, M.; De Vincenzi, A.; Silano, V. 2004. The new European legislation on traditional herbal medicines: main features and perspectives. **Fitoterapia 75:** 107-116.

Simoes, C.M.O.; Schenkel, E.P.; Gosmann, G.; Mello, J.C.P.; Mentz, L.A.; Petrovick, P.R. 2000. **Pharmacognosy: from the plant to the medicine**. Florianópolis. Ed. Universidade/UFRGS/Ed. Da UFSC. pp. 433-449.

Singh, B. and Usha, K. 2003. Salicylic acid induced physiological and biochemical changes in wheat seedlings under water stress. **Plant Growth Regulation 39:** 137-141.

Sirvent, T. and Gibson, D. 2002. Induction of hypericinsand hyperforin in *Hypericum perforatum* L. in response to biotic and chemical elicitors.

Physiological and Molecular Plant Biology 60: 311-320.

Shi, C.; Dai, Y.; Xia, B.; Xu, X.; Xie, Y.; Liu, Q. 2001. The purification and spectral properties of polyphenol oxidase I from *Nicotiniana tabacum*. **Plant Molecular Biology Reporter 19:** 318a-318h.

Sokmen, M.; Angelovab, M.; Krumova, E.; Pashova, S.; Ivancheva, S.; Sokmen, A.; Serkedjieva, J. 2005. In vitro antioxidant activity of polyphenol extracts with antiviralproperties from *Geranium sanguineum* L. **Life Sciences 76:** 2981-2993.

Staniszewska, I.; Królicka, A.; Malinski, E.; Lojkowska, E.; Szafranek, J. 2003. Elicitation of secondary metabolites in *in vitro* cultures of *Ammi majus* L. **Enzyme and Microbial Technology 33:** 565-568.

Sticher, L.; Mauch-Mani, B.; Métraux, J.P. 1997. Systemic acquired resistance. **Annual Review Phytopathology 35:** 235-270.

Taiz, L. and Zeiger, E. 2004. **Plant physiology**. Porto Alegre. Ed. Artmed. 719p.

Takeda, H.; Tsuji, M.; Inazu, M.; Egashira, T.; Matsumiya, T. 2002. Rosmarinic acid and caffeic acid produce antidepressant-like effect in the forced swimming test in mice. **European Journal of Pharmacology 449:** 261-267.

Tester, M. and Bacic, A. 2005. Abiotic stress tolerance in grasses. From model plants to crop plants. **Plant Physiology 137:** 791-793.

Valero, E. and Garcia-Carmona, F. 1992. Hysteresis and cooperative behavior of a latent plant polyphenol oxidase. **Plant Physiology 98:** 774-776.

Verpoorte, R. and Memelink, J. 2002. Engineering secondary metabolite production in plants. **Current Opinion in Biotechnology 13:** 181-187.

Wang, H.; Provan, G.J.; Helliwell, K. 2004. Determination of rosmarinic and caffeic acid in aromatic herbs by HPLC. **Food Chemistry 87:** 307-311.

Wang, L. and Zhang, H. 2005. A theoretical study of the different radicalscavenging activities of catechin, quercetin, and a rationally designed planar catechin. **Bioorganic Chemistry 33:** 108-115.

Wachowicz, C.M. and Carvalho, R.I.N. 2002. **Plant physiology: production and post-harvest.** Editora Champagnat: Curitiba, PR. 424p.

Wilt, F.M. and Miller, G.C. 1992. Seasonal variation of coumarin and flavonoid concentrations in persistent leaves of wyoming big sagebrush (*Artemisia tridentata ssp. wyomingensis* - Asteraceae). **Biochemical Systematics and Ecology 20** (1): 53-67.

Yamamoto, H.; Inoue, K.; Yazaki, K. 2000. Caffeic acid oligomers in *Lithospermum erythrorhizon* cell suspension cultures. **Phytochemistry 53:** 651-657.

Yao, H. and Tian, S. 2005. Effects of pre- and post-haverst application of salicylic acid or methyl jasmonate on inducing disease resistence of sweet cherry fruit in storage. **Postharvest Biology and Technology 35:** 253-262.

Y u, L-J.; Lan, W-Z.; Qin, W-M.; Xu, H-B. 2001. Effects of salicylic acid on fungal elicitor-induced membrane-lipid peroxidation and taxol production in cell suspension cultures of *Taxus chinensis*. **Process Biochemistry 37:** 477-482.

Y unes, R.A. and Calixto, J.B. 2001. **Medicinal Plants: from the perspective of Modern Medicinal Chemistry**. Chapecó. Ed. Argos. 523 p.

Y unes, R.A.; Pedrosa, R.C.; Cechinel F°, V. 2001. Drugs and Phytotherapics: The need to develop the Phytotherapics and Phytoceuticals industry in Brazil. **Quimica Nova 24** (1): 147-152.

Zheng, Z. and Shetty, K. 2000. Azo dye-mediated regulation of total phenolics and peroxidase activity in thyme (*Thymus vulgaris* L.) and rosemary (*Rosmarinus officinalis* L.) clonal lines. **Journal of Agricultural and Food Chemistry 48:** 932-937.

Printed by Books on Demand GmbH, Norderstedt / Germany